Arthur Stanley Eddington Centenary Lectures

EDDINGTON

The most distinguished astrophysicist of his time

Eddington by Augustus John

(Reproduced by permission of the Master and Fellows of Trinity College, Cambridge)

EDDINGTON

The most distinguished astrophysicist of his time

S. CHANDRASEKHAR

University of Chicago

CAMBRIDGE UNIVERSITY PRESS

Cambridge

London New York New Rochelle

Melbourne Sydney

CAMBRIDGE UNIVERSITY PRESS
Cambridge, New York, Melbourne, Madrid, Cape Town, Singapore,
São Paulo, Delhi, Dubai, Tokyo

Cambridge University Press
The Edinburgh Building, Cambridge CB2 8RU, UK

Published in the United States of America by Cambridge University Press, New York

www.cambridge.org
Information on this title: www.cambridge.org/9780521122429

First published 1983
This digitally printed version 2009

A catalogue record for this publication is available from the British Library

ISBN 978-0-521-25746-6 Hardback
ISBN 978-0-521-12242-9 Paperback

Contents

I

Eddington: The most distinguished astrophysicist of his time

May I begin by expressing my gratitude to the Master of Trinity and his Council for their trust in assigning to me the privilege of giving the Centenary Lectures in memory of one of the most distinguished members of the College and of the University. I knew Eddington as a member of the Fellowship of Trinity during the early and the middle thirties when, besides Eddington, it included J. J. Thomson, Ernest Rutherford, George Trevelyan, Douglas Adrian, Donald Robertson, G. H. Hardy, J. E. Littlewood, and a host of others. It is hardly necessary for me to say how much it means to me to have been a member of that society during those years and to be asked now, almost fifty years later, to give these lectures in honour of one whose personal friendship I was fortunate to enjoy.

I

When Eddington died in November 1944 at the age of sixty-two, Henry Norris Russell, his great contemporary across the Atlantic, wrote:[1] 'The death of Sir Arthur Eddington deprives astrophysics of its most distinguished representative.' I have taken my cue from Russell for the substance of this, the first of my two lectures.

Before I turn to an assessment of Eddington's contributions to astronomy and to astrophysics, I should like to start with a few biographical notes which may give some impression of the manner of man he was.

Arthur Stanley Eddington was born on December 20, 1882

at Kendal in Westmoreland. His father, Arthur Henry Eddington, was Headmaster and Proprietor of Stramongate School at Kendal where John Dalton had taught a century earlier. Forty-eight years later, when Eddington was conferred the Freedom of the Borough of Kendal, he recalled to say:[2]

The traditions of Kendal have been woven into my earliest memories as the home of the brief married life of my parents. I cannot but feel thankful that Kendal has recognised scientific work as a public service of the utmost importance, not in any material sense; but that it has contributed something to the community. Kendal has an earlier association with science, that great chemist, perhaps the greatest of all chemists, who was Headmaster of Stramongate School, the same School of which a century later my father became Headmaster and where I was born. From John Dalton we had the atom. Now I have become an atom chaser myself. John Dalton must have left some germ behind him which lingered in the walls of Stramongate. I like to think of that continuity, and I am proud to have been able in a way to follow in the path which has been opened out by Kendal's great scientist.

Eddington's father died in 1884; and his mother, together with her two young children, Stanley and his elder sister by four years, Winifred, moved to Weston-super-Mare. Here Eddington showed, quite early, his fascination with large numbers: he learnt the 24×24 multiplication table; and on one occasion started counting all the words in the Bible. Eddington never lost his fascination with large numbers. In later life, he often chose to write astronomical measures and distances with all their zeros explicitly. Thus, Eddington began an Evening Discourse to the British Association in Oxford in 1926 with:[3]

The stars are of remarkably uniform mass, that of the sun is – I will write it on the blackboard:

2 000 000 000 000 000 000 000 000 000 tons.

I hope I have counted the 0's rightly, though I dare say you would not mind if there were one or two too many or too few. But nature does *mind.*

And in 1935, when Eddington's interests had turned to the astronomical universe in the large, he introduced the subject with:[4]

	Miles
Distance of sun	93 000 000
Limit of solar system (Orbit of Pluto)	3 600 000 000
Distance of nearest star	25 000 000 000 000
Distance of nearest galaxy	6 000 000 000 000 000 000
Original circumference of the universe	40 000 000 000 000 000 000 000

With respect to large numbers, the most famous, of course, is the opening sentence[5] of Chapter XI of his *Philosophy of Physical Science* published in 1939:

> *I believe there are 15 747 724 136 275 002 577 605 653 961 181 555 468 044 717 914 527 116 709 366 231 425 076 185 631 031 296 protons in the universe, and the same number of electrons.*

This number, which is 136×2^{256}, has come to be known as Eddington's number. Bertrand Russell asked Eddington if he had computed this number himself or if he had someone else do it for him. Eddington replied that he had done it himself during an Atlantic crossing!

At Weston-super-Mare, Eddington attended Brymelyn School between 1893 and 1898. I recall Eddington telling me that at school one of the games he played was to make up English sentences which were grammatically correct but which made no sense, in the manner of Lewis Carroll. An example he gave me was

> *To stand by the hedge and sound like a turnip.*

In later life, Eddington was wont to introduce such sentences in his more serious writings to make a point. Thus, in his Swarthmore Lecture, *Science and the Unseen World*, we find:[6]

> *Human personalities are not measurable by symbols any more than you can extract the square root of a sonnet.*

I shall not go into any further details of his early education except to say that he entered Owen's College, Manchester, in

1898 and remained there for four years. His teachers in Owen's College included Sir Arthur Schuster and Sir Horace Lamb. Eddington seems to have maintained a deep admiration for Lamb all through his life. Thus, during the early twenties, when Eddington had become one of the most well-known figures of British science, he is reported to have said: 'While I know what it is to be treated something like a lion, I would rather like to become something of a Lamb.'

After a distinguished career at Manchester, Eddington proceeded to Cambridge in 1903 on a minor entrance-scholarship which was later changed to a major scholarship. He was the Senior Wrangler in 1904; was awarded the Smith's prize in 1907, and elected to a Fellowship in Trinity in the same year. In 1936, I had the occasion to sit next to Alfred North Whitehead at a dinner at Harvard University. Whitehead recalled that, as one of the electors in 1907, he had ensured the election of Eddington to the Fellowship in preference to another who had submitted a much more voluminous thesis. And Whitehead seemed proud to recall this fact.

In 1907, in the same year that he was elected to the Fellowship at Trinity, Eddington, at the invitation of Sir William Christie, the Astronomer Royal, joined the staff of the Greenwich Observatory as a Chief Assistant. He held this position for five years when in 1912 he was elected to the Plumian Chair in Cambridge as the successor to Sir George Darwin. And in 1914, at the death of Sir Robert Ball, Eddington became the Director of the Cambridge Observatory, as well. He held these positions with great distinction for the next thirty years. And in Cambridge he made his home, first with his mother and sister and later with his sister alone.

II

I shall conclude my biographical notes with some remarks on Eddington's general views and habits.

Eddington was a Quaker; and as a Quaker, he was a conscientious objector during the First World War. I shall have more to say about his conscientious objection to the War in my next lecture. Now, I shall refer only to his Swarthmore Lecture for 1929, *Science and the Unseen World.* In this Lecture, Eddington expresses with transparent sincerity his views on religion, science, and life generally. Let me read a few paragraphs which seem to summarize his views:[7]

> *Religious creeds are a great obstacle to any full sympathy between the outlook of the scientist and the outlook which religion is so often supposed to require . . . The spirit of seeking which animates us refuses to regard any kind of creed as its goal. It would be a shock to come across a university where it was the practice of the students to recite adherence to Newton's laws of motion, to Maxwell's equations and to the electromagnetic theory of light. We should not deplore it the less if our own pet theory happened to be included, or if the list were brought up to date every few years. We should say that the students cannot possibly realise the intention of scientific training if they are taught to look on these results as things to be recited and subscribed to. Science may fall short of its ideal, and although the peril scarcely takes this extreme form, it is not always easy, particularly in popular science, to maintain our stand against creed and dogma.*
>
> *Rejection of creed is not inconsistent with being possessed by a living belief. We have no creed in science, but we are not lukewarm in our beliefs. The belief is not that all the knowledge of the universe that we hold so enthusiastically will survive in the letter; but a sureness that we are on the road. If our so-called facts are changing shadows, they are shadows cast by the light of constant truth.*
>
> *There is a kind of sureness which is very different from cocksureness.*

On a lighter matter, it was well known among Eddington's friends that he greatly looked forward to his solitary cycling tours in the spring and in the fall. But, perhaps, only a few knew of the careful record he kept of them. Before I left Cambridge in December 1936, Eddington showed me a large Bartholomew Touring Map of England in which he had carefully traced out in black ink all the different routes he had taken over the years. He further told me that the map, that he had spread out before us, was the second one, that the first had been mutilated by his dog, and that he had to transcribe on a new map the many routes he had traced out on the earlier one!

Eddington also told me that, while he was a Chief Assistant at Greenwich, he and Sydney Chapman (another great cycling enthusiast) had devised a criterion for judging cycling records. The criterion was the largest number *n* such that one had cycled *n* or more miles on *n* different days. (At a later time, when I mentioned this criterion to Chapman, he had forgotten it; but he did remember that he and Eddington had often compared notes on their cycling tours.)

It is perhaps touching that in every letter that he wrote to me subsequently, he included his latest value of *n*. The following extracts are from two letters:

> *My cycling n is still 75. I was rather unlucky this Easter as I did two rides, seventy-four and three-quarters miles, which do not count; I still need four more rides for the next quantum jump. However, I had marvellously fine weather and splendid country, chiefly South Wales . . .*
>
> *. . . tomorrow I have to put on weird costume – knee breeches and silk hose! – and get my Order from the King.* (*1938 July 4*)
>
> *n is now 77. I think it was 75 when you were here. It made the last jump a few days ago when I took an eighty mile ride in the fen country. I have not been able to go on a cycling tour since 1940 because it is impossible to rely on obtaining accommodations for the night; so my records advance slowly.* (*1943 September 2*)

One last matter. Eddington was addicted to solving the crossword puzzles in *The Times* and in the *New Statesman and Nation*. He rarely took more than five minutes for a puzzle. On occasions, Eddington allowed me to watch him as he solved a puzzle; and I marvelled at his quickness.

III

I now turn to an assessment of Eddington's contributions to astronomy and to astrophysics.

At the time Eddington entered astronomy in 1906, a revolutionary discovery had been made by J. C. Kapteyn of Groningen. Kapteyn was a great pioneer in the study of stellar motions; and his discovery was the following.

It had been assumed till then that the motions of stars were entirely random, with no preferential direction, in a *local standard of rest* i.e., one in which the mean velocities of the stars, in the neighbourhood, is zero. And a basic problem in stellar motions, as revealed by the proper motions and the radial velocities of the stars, was to determine the *solar motion* i.e., the Sun's 'peculiar velocity' in the standard of rest determined by the stars in the neighbourhood of the Sun.

On the assumption of the randomness of the velocities with no direction preferentially distinguished, the distribution of the proper motions, projected on to a small region of the sky, must have the appearance of an elongated ellipse (see Fig. 1*a*). But this was not what Kapteyn found. He found, instead, a double-lobed curve as shown in Fig. 1*b*.

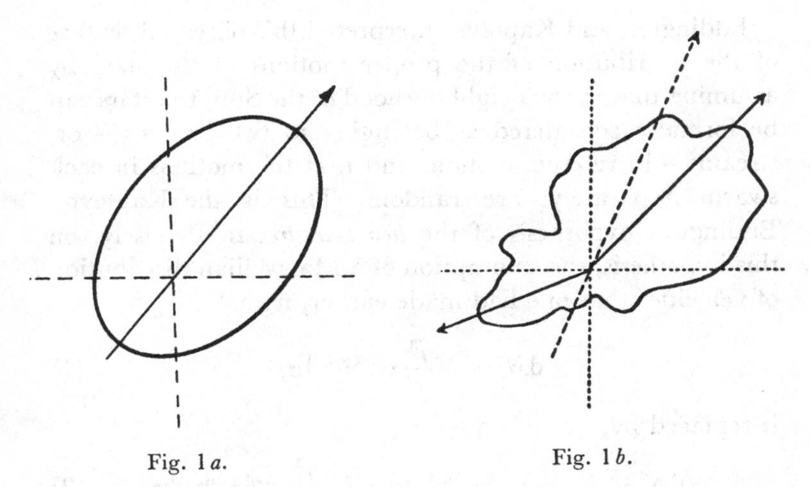

Fig. 1*a*. Fig. 1*b*.

Fig. 1.

Eddington has described Kapteyn's discovery as follows.[8]

There is now a long series of 'Groningen Publications' relating to these problems. The most interesting of them all is No. 6. But it is no use going to a library to consult it; for the interesting thing about No. 6 is that it was never written. Nature took an unexpected turn, and would not fit into the scheme which No. 6 was promised to elaborate. No. 5 was entitled 'The distribution of cosmic velocities: Part I. Theory'; it was a study of how the motions of stars pursuing their courses at random would turn out statistically when the solar motion and effects of varying distance were allowed for. Meanwhile the observed Auwers-Bradley proper motions were being prepared for comparison, so as to determine the numerical constants in the formulae. But the theory, though it represented the unquestioned views of the time, turned out to be so wide of the mark that not even the beginnings of a comparison were possible; and the application of the formulae had to be abandoned. This was Kapteyn's great discovery of the two star streams, announced at the British Association meeting in South Africa in 1905, which revealed for the first time a kind of organization in the system of the stars and started a new era in the study of the relationships of these widely separated individuals.

Eddington and Kapteyn interpreted this observed feature of the distribution of the proper motions of the stars by assuming that in the neighbourhood of the Sun, the stars can be formally considered as belonging to two swarms – or, streams – in relative motion, and that the motions in each swarm, separately, are random. This is the Kapteyn–Eddington hypothesis of the *two star-streams*. Precisely, on this hypothesis, the assumption of a Maxwellian distribution of velocities that one had made earlier, namely,

$$\mathrm{d}\mathcal{N} = \mathcal{N}\frac{j^3}{\pi^{\frac{3}{2}}}\,\mathrm{e}^{-j^2|\mathbf{u}|^2}\,\mathrm{d}\mathbf{u}, \tag{1}$$

is replaced by,

$$\mathrm{d}\mathcal{N} = \mathcal{N}_1\frac{j_1^3}{\pi^{\frac{3}{2}}}\,\mathrm{e}^{-j_1^2|\mathbf{u}-\mathbf{u}_1|^2}\,\mathrm{d}\mathbf{u} + \mathcal{N}_2\frac{j_2^3}{\pi^{\frac{3}{2}}}\,\mathrm{e}^{-j_2^2|\mathbf{u}-\mathbf{u}_2|}\,\mathrm{d}\mathbf{u}, \tag{2}$$

where $\mathcal{N}_1$ and $\mathcal{N}_2$ are the numbers of stars in the two streams, $\mathbf{u}_1$ and $\mathbf{u}_2$ are the velocities of the two streams in the local standard of rest, and j_1 and j_2 are the reciprocal mean speeds of the stars in the two streams, separately. Further the con-

Fig. 2
A. S. Eddington and J. C. Kapteyn
(Rome, May, 1922)

dition, that the distribution of the velocities (2) is with respect to the local standard of rest, requires that

$$\mathcal{N}_1 \mathbf{u}_1 + \mathcal{N}_2 \mathbf{u}_2 = 0. \quad (3)$$

Eddington wrote several papers during his Greenwich years on the kinematics of the motions described by the distribution (2). He also developed analytic methods for determining the parameters, $\mathcal{N}_1$, $\mathcal{N}_2$, $\mathbf{u}_1$, $\mathbf{u}_2$, j_1, and j_2 of the two streams. And he determined them from the proper motions of the stars that were available to him. These papers of Eddington represent a remarkable synthesis of theory and observations; and they reveal his discriminating insight into the data of astronomical observations.

Adequate as the Kapteyn–Eddington hypothesis of the two star-streams was, an alternative interpretation of the same observations was given by Karl Schwarzschild. Schwarzschild's starting point was to replace the Maxwellian distribution (1) by the more general *ellipsoidal distribution*,

$$d\mathcal{N} = \mathcal{N}\frac{j_1 j_2 j_3}{\pi^{\frac{3}{2}}} \exp\left(-j_1^2 u_1^2 - j_2^2 u_2^2 - j_3^2 u_3^2\right) du_1 du_2 du_3, \quad (4)$$

in a suitably oriented frame of reference. Eddington has, himself, described Schwarzschild's alternative formulation as the most elegant and the most satisfying way of interpreting Kapteyn's discovery; and this interpretation has survived to this day.

This phase of Eddington's contributions came to an end with the publication in 1914 of his first book, *Stellar Movements and the Structure of the Universe*. This book, in large part, is devoted to a systematic presentation of the then extant knowledge of stellar motions. But the last chapter, 'On the Dynamics of Stellar Systems', breaks new and fertile ground. After showing that binary stellar encounters cannot be effective in changing the directions of motions of the individual stars, Eddington concluded that the function $f(x, y, z, u, v, w, t)$ governing the distribution of the stars in the six-dimensional phase-space, must be determined by the dynamical orbits described by the stars in the smoothed-out gravitational potential of the whole system i.e., by a solution of the six-dimensional Liouville equation (or the collisionless Boltzmann-equation as we would now call it).

In papers published in 1915 and 1916, Eddington sought solutions of Liouville's equation which are consistent with

Schwarzschild's ellipsoidal distribution of velocities; and he obtained, in particular, a self-consistent solution appropriate for spherically symmetric stellar systems. In this same general context, Eddington pointed out, for the first time, how the virial theorem can be used to relate the average kinetic energy of a star in a cluster with its mean potential energy – a method which continues to be in vogue in the still larger context of galaxies and clusters of galaxies.

By these investigations, Eddington may be said to have founded the subject of stellar dynamics – a discipline which now stands on its own right.

IV

I now turn to what are, undoubtedly, Eddington's most significant contributions to the physical sciences: his founding of modern theoretical astrophysics and his creating the discipline of the structure, the constitution, and the evolution of the stars. His interest in the constitution of the stars was stimulated in 1916 by his efforts to understand the nature of Cepheid variability; and it culminated in 1926 with the publication of his *Internal Constitution of the Stars*. It should be remembered that during this same ten-year period, he was involved in the eclipse expeditions which provided the first verification of the deflection of light in its passage through a gravitational field – an event which he was later to describe as the most exciting event in his connection with astronomy – and in the writing of his *Mathematical Theory of Relativity* (1923) not to mention his *Report on the Relativity Theory of Gravitation* (1915) and his two popular books, *Space, Time, and Gravitation* (1920) and *Stars and Atoms* (1927) – a ten-year period of remarkable productivity.

In the domain of the internal constitution of the stars, Eddington recognized and established the following basic elements of our present understanding:

1. Radiation pressure must play an increasingly important role in maintaining the equilibrium of stars of increasing mass.

2. In parts of the star in which radiative equilibrium, as distinct from convective equilibrium, obtains, the temperature gradient is determined jointly by the distribution of the energy sources and of the opacity of the matter to the prevailing radiation field. Precisely,

$$\frac{dp_r}{dr} = -\kappa \frac{L(r)}{4\pi c r^2}\rho, \quad p_r = \tfrac{1}{3}aT^4, \tag{5}$$

and

$$L(r) = 4\pi \int_0^r \epsilon \rho r^2 dr, \tag{6}$$

where p_r, κ, ϵ, and ρ denote, respectively, the radiation pressure, the coefficient of stellar opacity, the rate of energy generation per gram of the stellar material, and the density. Also, a is Stefan's radiation constant and c is the velocity of light.

3. The principal physical process contributing to the opacity, κ, is determined by the photo-electric absorption coefficient in the soft X-ray region i.e., by the ionization of the innermost K- and L-shells of the highly ionized atoms.

4. With electron scattering as the ultimate source of stellar opacity, there is an upper limit to the luminosity, L, that can support a given mass M. The maximum luminosity, set by the inequality,

$$L < \frac{4\pi c G M}{\sigma_e}, \tag{7}$$

where σ_e denotes the Thomson scattering-coefficient, is now generally referred to as the *Eddington limit.* This Eddington limit plays an important role in current investigations relating to X-ray sources and the luminosity of accretion discs around black holes.

5. In a first approximation, in normal stars (i.e., in stars along the main sequence) the (mass, luminosity, effective temperature)-relation is not very sensitive to the distribution of the energy sources through the star. Therefore, a relation is available for comparison with observations even in the absence of a detailed knowledge of the energy sources of the star.

6. The burning of hydrogen into helium is the most likely source of stellar energy.

These deductions, which Eddington made some sixty years ago, continue to be valid – then, as now.

I should like to expand on some of these deductions for illustrating Eddington's approach to problems of this kind.

I shall consider, first, the way Eddington established the increasing importance of radiation pressure as a factor in the equilibrium of stars of increasing mass. It will be recalled that when Eddington wrote out the mass of the Sun in tons, with all the zeros explicitly, he remarked that one should not consider the exact number of zeros as of no particular consequence since 'nature *does* mind'.

In his *Internal Constitution of the Stars* (p. 16), Eddington arrives at his conclusion by imagining a physicist[9]

> *on a cloud-bound planet, who has never heard tell of the stars, calculating the ratio of radiation pressure to gas pressure for a series of globes of gas of various sizes, starting, say, with a globe of mass 10 g, then 100 g, 1000 g and so on, so that his n-th globe contains 10^n g. [Table 1] shows the more interesting part of his results.*

Table 1

No. of Globe	Radiation Pressure	Gas Pressure
30	0.00000016	0.99999984
31	0.000016	0.999984
32	0.0016	0.9984
33	0.106	0.894
34	0.570	0.430
35	0.850	0.150
36	0.951	0.049
37	0.984	0.016
38	0.9951	0.0049
39	0.9984	0.0016
40	0.99951	0.00049

> *The rest of the table would consist mainly of long strings of 9's and 0's. Just for the particular range of mass about the 33rd to*

35th globes the table becomes interesting, and then lapses back into 9's and 0's again. Regarded as a tussle between matter and aether (gas pressure and radiation pressure) the contest is overwhelmingly one-sided except between numbers 33–35, where we may expect something to happen.

What 'happens' is the stars.

We draw aside the veil of cloud beneath which our physicist has been working and let him look up at the sky. There he will find a thousand million globes of gas nearly all of mass between his 33rd and 35th globes – that is to say, between ½ and 50 times the Sun's mass. The lightest known star is about $3 \cdot 10^{32}$ g and the heaviest about $2 \cdot 10^{35}$ g. The majority are between 10^{33} and 10^{34} g, where the serious challenge of radiation pressure to compete with gas pressure is beginning.

The calculations of Table 1 are made on the assumption that the ratio of the gas pressure (p_g) to the radiation pressure (p_r), $\beta/(1-\beta)$, is constant through the star; and that the mean molecular weight, μ, has the value 2.5. A value of $\mu = 1.0$ would have been more realistic, in which case the mass of each of the globes would have to be increased by a factor $(2.5)^2 = 6.25$. This factor is not of much consequence. But there are two important aspects of the argument on which Eddington is silent. One is that while a combination of the natural constants of the dimensions of a mass and of stellar magnitude (with all the zeros!) is clearly implied by the calculations, Eddington does not isolate it – a surprising omission in view of his later preoccupations with natural constants. Actually, the combination of natural constants which determines the masses of the globes, in the interesting range, is

$$\left[\left(\frac{k}{H}\right)^4 \frac{3}{a}\right]^{\frac{1}{2}} \frac{1}{G^{\frac{3}{2}}}, \tag{8}$$

where H is the mass of the proton, G is the constant of gravitation, and k and a are the Boltzmann and the Stefan constants, respectively. On inserting for Stefan's constant its value,

$$a = \frac{8\pi^5}{15} \frac{k^4}{h^3 c^3}, \tag{9}$$

where h is Planck's constant, we find that the combination of the natural constants, of the dimensions of a mass, that is involved is

$$\left(\frac{hc}{G}\right)^{\frac{3}{2}} \frac{1}{H^2} \simeq 29.2 \odot \simeq 5.2 \times 10^{34} \text{ g.} \tag{10}$$

A case can be made that the successes of the current theories of stellar structure and stellar evolution derive in large measure from the foregoing combination of the natural constants providing a mass of the correct stellar magnitude. (I shall note in passing that the more general combination of the dimensions of a mass, namely

$$\left(\frac{hc}{G}\right)^{\alpha} \frac{1}{H^{2\alpha-1}}, \tag{11}$$

where α is arbitrary, includes the Planck mass $(hc/G)^{\frac{1}{2}}$, when $\alpha = \frac{1}{2}$.)

A second aspect of the calculations, presented in Table 1, on which Eddington is also silent, is: Why is the extent to which radiation pressure provides support against gravity relevant to the "happening of the stars"? On this question, Eddington, instead of basing his arguments on his 'standard model' (in which the ratio of the radiation pressure to the gas pressure is a constant through the star), could have used a theorem of his own – namely, that for *stability*, the pressure at the centre of a star must be less than that at the centre of a configuration of uniform density of the same mass and the same central density – to show that the ratio of the radiation pressure to the total pressure at the centre of a star must be less than a certain fraction dependent on the mass of the star only. Table 1 would then be replaced by Table 2; and the conclusion of the physicist on the cloud-bound planet would have been the same.

With regard to the other key points, the one relating to the source of stellar opacity reveals best the manner of Eddington's acceptance of physical theories when confronted with his astronomical deductions. In his considerations relating to the problem of stellar opacity, he had the benefit of consultations with C. D. Ellis who was an expert in X-ray and γ-ray

Table 2

$(M/\odot)\mu^2$	Radiation Pressure	Gas Pressure
0.56	0.01	0.99
1.01	0.03	0.97
2.14	0.10	0.90
3.83	0.20	0.80
6.12	0.30	0.70
9.62	0.40	0.60
15.49	0.50	0.50
26.52	0.60	0.40
50.92	0.70	0.30
122.5	0.80	0.20
224.4	0.85	0.15
519.6	0.90	0.10

physics. Also, a famous paper by H. A. Kramers provided the first theoretical evaluation of the atomic cross-sections for photo-electric ionization which Eddington needed. But prior to the publication of Kramers's paper, Eddington had developed a theory of his own based on the physically untenable hypothesis of the direct capture of electrons by atomic nuclei. And he maintained the astronomical relevance of his theory against cogent arguments by many physicists including Rutherford. And he abandoned his theory only after the appearance of Kramers's paper and the agreement of the theoretical calculations with experimental results had been demonstrated. But the comparison of astronomical observations with the theoretical mass–luminosity relation, adopting Kramers's opacity law, left a discrepancy of a factor exceeding ten. Eddington (and, independently, Strömgren) eliminated this discrepancy in 1932 by adopting the large abundances of hydrogen and helium which had, by then, been established by Russell from an analysis of the abundances of the elements in the solar atmosphere. Eddington was however slow to accept the non-universality of the composition of the stars that followed. Eddington had in fact realized much earlier that the assumption that hydrogen was abundant and that a mean molecular weight of the stars close to unity would resolve the opacity discrepancy. But this assumption would have spoiled his argument for the 'happening of the stars'; and it is characteristic of Eddington that he should have concluded 'I would much prefer to find some other explanation for the discordance'.[9]

Among Eddington's predictions, that of the source of stellar energy is perhaps the most spectacular. His address, on August 24, 1920 to the British Association meeting in Cardiff, contains some of the most prescient statements in all of astronomical literature.[10]

> *Only the inertia of tradition keeps the contraction hypothesis alive – or rather, not alive, but an unburied corpse. But if we decide to inter the corpse, let us frankly recognize the position in which we are left. A star is drawing on some vast reservoir of energy by means unknown to us. This reservoir can scarcely be other than the sub-atomic energy which, it is known, exists abundantly in all matter; we sometimes dream that man will one day learn how to release it and use it for his service. The store is well-nigh inexhaustible, if only it could be tapped. There is sufficient in the Sun to maintain its output of heat for 15 billion years . . .*
>
> *Aston has further shown conclusively that the mass of the helium atom is even less than the sum of the masses of the four hydrogen atoms which enter into it – and in this, at any rate, the chemists agree with him. There is a loss of mass in the synthesis amounting to 1 part in 120, the atomic weight of hydrogen being 1.008 and that of helium just 4. I will not dwell on his beautiful proof of this, as you will no doubt be able to hear it from himself. Now mass cannot be annihilated, and the deficit can only represent the mass of the electrical energy set free in the transmutation. We can therefore at once calculate the quantity of energy liberated when helium is made out of hydrogen. If 5 per cent of a star's mass consists initially of hydrogen atoms, which are gradually being combined to form more complex elements, the total heat liberated will more than suffice for our demands, and we need look no further for the source of a star's energy.*
>
> *If, indeed, the sub-atomic energy in the stars is being freely used to maintain their great furnaces, it seems to bring a little nearer to fulfilment our dream of controlling this latent power for the well-being of the human race – or for its suicide.*

In this context of the source of stellar energy, astrophysicists of an earlier generation will recall Eddington's famous retort:[11]

> *It has, for example, been objected that the temperature of the stars is not great enough for the transmutation of hydrogen into helium – so ruling out one possible source of energy. But helium exists, and it is not much use for the critic to urge that the stars are not hot enough for its formation unless he is prepared to show us a hotter place.*

V

I stated at the outset that Eddington's interest in the internal constitution of the stars arose from his efforts to find an explanation for stellar variability and the period–luminosity relation exhibited by the Cepheids. Eddington first generalized Ritter's earlier analysis of the adiabatic pulsations of gaseous stars in convective equilibrium to the case of a star in radiative equilibrium built on his standard model. Then combining the resulting formula for the period with his mass–luminosity relation, Eddington was able to account, in a general way, for the observed period–luminosity relation of the Cepheids. The pulsation theory of stellar variability thus came to be established.

Eddington's preliminary analysis of Cepheid variability did not provide the correct phase relationships among the various variables such as the brightness, the effective temperature, and the radial velocity of the star. However, he clearly realized that these phase relationships can be understood only by a careful examination of the mechanism of energy transfer in the outer layers of the star where the abundant elements, hydrogen and helium, get ionized and zones of convection are formed. He returned to this problem several times in later years. Indeed, one of his last published papers is devoted to this problem. But the final solution was found only later by the combined investigations of M. Schwarzschild, P. Ledoux, and R. Christy.

While Eddington's major contributions to astrophysics lay

in the domain of stellar structure, his contributions to other areas of astrophysics are by no means insignificant. He devised a method of approximation – the 'Eddington approximation' – for solving problems in radiative transfer. His solution for the problem of line formation in stellar atmospheres, for example, was much in use during the pioneering years in the theory of stellar atmospheres. He also considered the effect of reflexion in close binaries – an effect which one must allow for in analysing the light curves of eclipsing binaries for the purposes of determining the masses of the components. The problem Eddington considered in this latter context, is the prototype of the larger problem of the diffuse reflexion and transmission of light by plane-parallel atmospheres – a subject which grew to maturity in later years.

In these other areas of astrophysics, perhaps the most important is Eddington's introduction of a '*dilution factor*' – a word he coined and still in current usage – to allow for the reduced intensity of the prevailing radiation-field in determining the state of ionization in interstellar space. Eddington was also the first to adapt to the problem of *interstellar* absorption-lines the method of the 'curve of growth' which A. Unsold and M. Minnaert had developed for determining the relative abundances of the elements from the intensities of *stellar* absorption-lines.

Eddington's interests in galactic dynamics and astrophysics converged in his prediction that the radial velocities, determined from interstellar absorption-lines, must show, in their dependence on galactic latitude, an amplitude which is one-half of that shown by the stellar absorption-lines. This prediction was later beautifully confirmed by the observations of O. Struve and J. S. Plaskett.

Of the *Internal Constitution of the Stars*, which includes most of what I have described so far, Russell has said[12]

> *This volume has every claim to be regarded as a masterpiece of the first rank.*

VI

In my account of Eddington's contributions to the constitution of the stars, I did not make any reference to the running controversies he had, first with Jeans and later with Milne. From the vantage point of the present, the questions that were at issue then do not seem to be very relevant: with the understanding of the source of stellar energy, the unresolved issues have required different formulations and different solutions. It should, however, be stated that Eddington was not always fair in the way he treated his scientific adversaries. For example, after a paper by Milne presented at the December 1929 meeting of the Royal Astronomical Society, Eddington's response, in part, was[13]

> *Prof. Milne did not enter into detail as to why he arrives at results so widely different from my own; and my interest in the rest of the paper is dimmed because it would be absurd to pretend that I think there is the remotest chance of his being right.*

And the acrimony with which the discussions were sometimes carried out is sufficiently illustrated by the following extracts from two letters of Jeans published in the *Observatory*:[14]

> *So much work has been done on isothermal equilibrium that it is difficult to understand how Prof. Eddington can harbour the illusion that he is doing pioneer work in unexplored territory, yet his complete absence of reference to other theoretical workers (except for some numerical computations quoted from Emden) suggests that such is actually the case. (August 1926)*

> *May I conclude by assuring Prof. Eddington it would give me great pleasure if he could remove a long-standing source of friction between us by abstaining in future from making wild attacks on my work which he cannot substantiate, and by making the usual acknowledgements whenever he finds that my previous work is of use to him? I attach all the more importance to the second part of the request, because I find that some of the most fruitful ideas which I have introduced into astronomical physics – e.g., the*

annihilation of matter as a source of stellar energy, and highly dissociated atoms and free electrons as the substance of the stars – are by now fairly generally attributed to Prof. Eddington. (November 1926)

I should like to leave these unhappy episodes with an anecdote in a lighter vein.

It is known that Eddington, on occasions, enjoyed seeing horse races and that, not infrequently, took his sister to the Newmarket races. G. H. Hardy must have known of this, for I heard him once ask Eddington if he had ever bet on a horse. Eddington admitted that he had, 'but just once'. Hardy was curious to know the occasion; and Eddington explained that a horse named Jeans was running and that he could not resist the temptation of betting on it. Questioned whether he had won, Eddington responded with his characteristic smile and a 'NO!'

2

Eddington: The expositor and the exponent of general relativity

My last lecture was devoted mostly to Eddington's contributions to theoretical astrophysics and to justifying Russell's assessment of him as the most distinguished representative of astrophysics of his time. In this lecture, I shall turn to Eddington as an expositor and an exponent of the general theory of relativity, to the part he played in the Greenwich–Cambridge expeditions to observe the solar eclipse of May 29, 1919 with the express purpose of verifying Einstein's prediction of the deflection of light by a gravitational field, and to his efforts, extending over sixteen years, in cosmology and – quoting his own description – in 'unifying quantum theory and relativity theory'. But in contrast to my last lecture, I am afraid that this lecture will not altogether be a happy one.

I shall begin with the happier side.

I

After founding the principles of the special theory of relativity in 1905, Einstein's principal preoccupation in the ten following years was to bring the Newtonian theory of gravitation into conformity with those same principles and, in particular, with the requirement that no signal be propagated with a velocity exceeding that of light. After many false starts, Einstein achieved his goal in a spectacular series of short

Fig. 3

A. Einstein, P. Ehrenfest, W. deSitter
A. S. Eddington, H. A. Lorentz
(Leiden, deSitter's Study, September 26, 1923)

communications to the Berlin Academy of Sciences during the summer and the autumn of 1915. Because of the war, the news of Einstein's success would not have crossed the English Channel (not to mention the Atlantic Ocean) had it not been for the neutrality of the Netherlands and Einstein's personal friendship with Lorentz, Ehrenfest, and deSitter. deSitter sent copies of Einstein's papers to Eddington; and he further communicated to the Royal Astronomical Society, during

1916–17, three papers of his own in part expository and in part original contributions. What has now come to be called deSitter's universe is described in the last of the three papers.

Eddington, as the Secretary of the Royal Astronomical Society at that time, had to deal with deSitter's papers. One may presume, from his account[15] of the last of the three papers at the December, 1917 meeting of the Royal Astronomical Society, that he had read the papers carefully and refereed them himself.

It will be recalled that in the last of the communications in which Einstein formulated his fundamental field equations, he concluded with the prophetic statement, 'scarcely anyone who has fully understood this theory can escape from its magic'. Eddington must surely have been caught in its magic; for, within two years, he had written his *Report on the Relativity Theory of Gravitation* for the Physical Society of London, a report that must have been written in white heat. Eddington's *Report* is written so clearly and yet so concisely that it can be read, even today, as a good introductory text by a beginning student.

II

Eddington's enthusiasm for the general theory of relativity must have succeeded in ensnaring into its magic his close friend and associate Sir Frank Dyson, the Astronomer Royal; for together, they were soon planning expeditions to observe the solar eclipse of May 29, 1919 if, to quote Jeans, 'the state of civilization should permit when the time came'. Eddington has described his own part in the planning and in the successful outcome of the expeditions 'as the most exciting event, I recall, in my connection with astronomy'. The story has so many interesting facets that it is hard to know where to begin. I hope you will forgive me if I begin with an account which Eddington gave me.

I once expressed to Eddington my admiration of his scientific sensibility in planning the expeditions under circumstances when the future must have appeared very bleak. To my surprise, Eddington disclaimed any credit on that account and told me that, had he been left to himself, he would not have planned the expeditions since he was fully convinced of the truth of the general theory of relativity! And he told me how his part in the expeditions came about. I have written about it in the *Notes and Records of the Royal Society*[16]; but perhaps you will allow me to repeat it.

In 1917, after more than two years of war, England enacted conscription for all able-bodied men; and Eddington, who was then thirty-four, was eligible for draft. But as a practising and devout Quaker, he was a conscientious objector; and it was known and expected that he would claim deferment from military service on that account. Now the climate of opinion in England during World War I was very adverse with respect to conscientious objectors: it was in fact a social disgrace to be associated with one. And the stalwarts of the Cambridge of those days, Sir Joseph Larmor (of the Larmor precession), Professor H. F. Newall, and others tried through the Home Office to have Eddington deferred on the grounds that he was a most distinguished scientist and that it was not in Britain's long-range interests to have Eddington serve in the army. The case of Moseley killed in action at Gallipoli was very much in the minds of British scientists. And Larmor and others very nearly succeeded in their efforts. A letter from the Home Office was sent to Eddington, and all he had to do was to sign his name and return it. But Eddington added a postscript to the effect that if he were not deferred on the stated ground, he would claim it on grounds of conscientious objection anyway. This postscript naturally placed the Home Office in a logical quandary since a confessed conscientious objector must be sent to a camp; and Larmor and others were very much piqued. But as Eddington told me, he could see no reason for their pique. As he expressed himself, many of his Quaker friends found themselves in camps in Northern England peeling potatoes, and he saw no reason why he should not join them. In any event, apparently at Dyson's intervention – as the Astronomer Royal he had close connections with the Admiralty – Eddington was deferred with the express stipulation that if the war should end by May 1919, then Eddington should undertake to lead an expedition for the purpose of verifying Einstein's prediction!

Slightly different accounts of these incidents have been published; but they differ only in the emphasis and in the overtones. In any event, we are fortunate in having an account of the planning and the execution of the expeditions by Eddington himself. He writes:[17a, 17b]

The bending affects stars seen near the sun, and accordingly the only chance of making the observation is during a total eclipse when the moon cuts off the dazzling light. Even then there is a great deal of light from the sun's corona which stretches far above the disc. It is thus necessary to have rather bright stars near the sun, which will not be lost in the glare of the corona. Further, the displacements of these stars can only be measured relatively to other stars, preferably more distant from the sun and less displaced; we need therefore a reasonable number of outer bright stars to serve as reference points.

In a superstitious age a natural philosopher wishing to perform an important experiment would consult an astrologer to ascertain an auspicious moment for the trial. With better reason, an astronomer today consulting the stars would announce that the most favourable day of the year for weighing light is May 29. The reason is that the sun in its annual journey round the ecliptic goes through fields of stars of varying richness, but on May 29 it is in the midst of a quite exceptional patch of bright stars – part of Hyades – by far the best star-field encountered. Now if this problem had been put forward at some other period of history, it might have been necessary to wait some thousands of years for a total eclipse of the sun to happen on the lucky date. But by strange good fortune an eclipse did happen on May 29, 1919 . . .

Attention was called to this remarkable opportunity by the Astronomer Royal (Sir Frank Dyson) in March 1917; and preparations were begun by a committee of the Royal Society and Royal Astronomical Society for making the observations . . .

. . . Plans were begun in 1918 during the war, and it was doubtful until the eleventh hour whether there would be any possibility of the expeditions starting . . . Two expeditions were organized at Greenwich by Sir Frank Dyson, the one going to Sobral in Brazil and the other to the Isle of Principe in West

Africa. Dr. A. C. D. Crommelin and Mr. C. Davidson went to Sobral; and Mr. E. T. Cottingham and the writer went to Principe.

It was impossible to get any work done by instrument-makers until after the armistice; and, as the expeditions had to sail in February, there was a tremendous rush of preparation. The Brazil party had perfect weather for the eclipse; through incidental circumstances, their observations could not be reduced until some months later, but in the end they provided the most conclusive confirmation. I was at Principe. There the eclipse day came with rain and cloud-covered sky, which almost took away all hope. Near totality the sun began to show dimly; and we carried through the programme hoping that the conditions might not be so bad as they seemed. The cloud must have thinned before the end of totality, because amid many failures we obtained two plates showing the desired star-images. These were compared with plates already taken of the same star-field at a time when the sun was elsewhere, so that the difference indicated the apparent displacement of the stars due to the bending of the light-rays in passing near the sun.

As the problem then presented itself to us, there were three possibilities. There might be no deflection at all; that is to say, light might not be subject to gravitation. There might be a 'half-deflection', signifying that light was subject to gravitation, as Newton had suggested, and obeyed the simple Newtonian law. Or there might be a 'full-deflection', confirming Einstein's instead of Newton's law. I remember Dyson explaining all this to my companion Cottingham, who gathered the main idea that the bigger the result, the more exciting it would be. 'What will it mean if we get double the deflection?' 'Then,' said Dyson, 'Eddington will go mad, and you will have to come home alone.'

Arrangements had been made to measure the plates on the spot, not entirely from impatience, but as a precaution against mishap on the way home, so one of the successful plates was examined immediately. The quantity to be looked for was large as astronomical measures go, so that one plate would virtually decide the question, though, of course, confirmation from others would be sought. Three days after the eclipse, as the last lines of the calculations were reached, I knew that Einstein's theory had stood the test and the new outlook of scientific thought must prevail. Cottingham did not have to go home alone.

The scientific results of the expedition were reported at a joint meeting of the Royal Society and the Royal Astronomical Society on November 6, 1919 with Sir J. J. Thomson, President of the Royal Society, in the Chair. This meeting was surrounded by an unusual amount of publicity; the extent of it was recalled by Rutherford on an occasion that I well remember.

It was during the Christmas recess of 1933 when, after dinner in Hall at Trinity, Rutherford, Eddington, Patrick DuVal (a distinguished geometer), Sir Maurice Amos (at one time, during the 1920s, the Chief Judicial Advisor to the Egyptian Government), and I sat around the fire in the Senior Combination Room, in conversation. At some point during the conversation, Sir Maurice Amos turned to Rutherford and said, 'I do not see why Einstein is accorded a greater public acclaim than you. After all, you invented the nuclear model of the atom; and that model provides the basis for all of physical science today and it is even more universal in its applications than Newton's laws of gravitation. Also, Einstein's predictions refer to such minute departures from the Newtonian theory that I do not see what all the fuss is about.' Rutherford, in response, turned to Eddington and said, 'You are responsible for Einstein's fame'. And more seriously, he continued:

The war had just ended; and the complacency of the Victorian and the Edwardian times had been shattered. The people felt that all their values and all their ideals had lost their bearings. Now, suddenly, they learnt that an astronomical prediction by a German scientist had been confirmed by expeditions to Brazil and West Africa and, indeed, prepared for already during the war, by British astronomers. Astronomy had always appealed to public imagination; and an astronomical discovery, transcending worldly strife, struck a responsive chord. The meeting of the Royal Society, at which the results of the British expeditions were reported, was headlined in all the British papers: and the typhoon of publicity crossed the Atlantic. From that point on, the American press played Einstein to the maximum.

I will conclude this account by quoting Jeans[18] when he presented the Gold Medal of the Royal Astronomical Society to Dyson:

In 1918, in the darkest days of the war, two expeditions were planned, one by Greenwich Observatory and one by Cambridge, to observe, if the state of civilization should permit when the time came, the eclipse of May 1919, with a view to a crucial test of Einstein's generalized relativity. The armistice was signed in November 1918; the expeditions went, and returned bringing back news which changed, and that irrevocably, the astronomer's conception of the nature of gravitation and the ordinary man's conception of the nature of the universe in which he lives. If the credit of this achievement had to be divided between Sir Frank Dyson and Professor Eddington I frankly do not know in what proportion the division should be made. To my mind, however, it is not so much an occasion for sharing out credit as for attributing the whole credit to each, for if either had failed to play his part, either from want of vision, of enthusiasm, or of capacity of seizing the right moment, I doubt if the expeditions would have gone at all, and the great credit of first determining observationally what sort of things space and time really are would probably have gone elsewhere.

III

I should like to add a few footnotes to the story.

At the conclusion of the reports by Dyson and by Eddington on the results of their expeditions, J. J. Thomson made the following remarks[19] from the Chair:

Newton did, in fact, suggest this very point in the first query in his 'Optics', and his suggestion would presumably have led to the half-value. But this result is not an isolated one; it is part of a whole continent of scientific ideas affecting the most fundamental concepts of physics . . . This is the most important result obtained in connection with the theory of gravitation since Newton's day, and it is fitting that it should be announced at a meeting of the Society *so closely connected with him . . .*

If his theory is right, it makes us take an entirely new view of gravitation. If it is sustained that Einstein's reasoning holds good – and it has survived two very severe tests in connection with the perihelion of Mercury and the present eclipse – then it is the result of one of the highest achievements of human thought. The weak point in the theory is the great difficulty in expressing it. It would seem that no one can understand the new law of gravitation without a thorough knowledge of the theory of invariants and of the calculus of variations.

The 'difficulty' of understanding the general theory of relativity, to which J. J. Thomson referred, was then, and for a long time, a prevalent view. Indeed, a myth soon arose that 'only three persons in the world understood general relativity'. The myth, in fact, had its origin at this same meeting.

Eddington recalled (during the after-dinner conversation at Trinity to which I referred earlier) that, as the joint meeting of the Royal Society and the Royal Astronomical Society was dispersing, Ludwig Silberstein came up to him and said, 'Professor Eddington, you must be one of three persons in the world who understands general relativity.' On Eddington's demurring to this statement, Silberstein responded 'Don't be modest, Eddington,' and Eddington replied that 'On the contrary, I am trying to think who the third person is.'

I may parenthetically remark that this supposed difficulty in understanding the general theory of relativity was greatly exaggerated: it contributed to the stagnation of the subject for several decades. Many of the developments of the sixties and the seventies could easily have taken place during the twenties and the thirties.

Eddington was very fond of repeating Dyson's remark to Cottingham, 'Eddington will go mad and you will have to come home alone.' Thus when at a meeting of the Royal Astronomical Society in January 1932, Finlay Freundlich reported [20] that the results of *his* eclipse expedition gave for the deflection of light a value substantially in excess of Einstein's prediction, Eddington repeated Dyson's remark with devastating effect! But when the concordant results of the Lick-Observatory expedition of 1922 were reported at the April 1923 meeting of the Royal Astronomical Society, Eddington's comment was[21]

> *I think that it was Bellman in 'The Hunting of the Snark' who laid down the rule 'When I say it three times, it is right'. The stars have now said it three times to three separate expeditions; and I am convinced their answer is right.*

And finally, I want to refer to a problem in probability theory which attained some notoriety in the late thirties and which had its origin in the Greenwich–Cambridge expeditions. The problem and the manner of its solution by Eddington reveal his insight into probability theory – a theory in which (as '*Combination of Observations*') he had become interested when he had to combine a large body of uncertain observations in his investigations in star streaming.

You will recall that the two eclipse expeditions were in charge, respectively, of Crommelin and Davidson (who went to Sobral) and Cottingham and Eddington (who went to Principe). In an after-dinner speech, before the expeditions departed, Crommelin hinted that the following situation might arise:

If C, C′, D and E each speak the truth once in three times, independently, and C affirms that C′ denies that D declares that E is a liar, what is the probability that E was speaking the truth?

Eddington stated this problem with A, B, C, and D substituted for C, C′, D, and E in his *New Pathways in Science*[22] and gave the solution as 25/71. He later stated that 'when I rashly gave my solution, I did not foresee a considerable increase in my correspondence'. Many, including Dingle, argued that the problem, as formulated, was ambiguous and that Eddington's solution was not, by any means, the most obvious one. Eddington meant the problem to assert (as anyone unsophisticated will agree) that two statements had been made:[23]

(1) D made a statement, say, X;

and

(2) A made the statement that 'B denies that C contradicted X'.

What is required is the probability that X is true. Eddington explained his solution as follows:*

We do not know that B and C made any relevant statements. For example, if B truthfully denied that C contradicted X, there is no reason to suppose that C affirmed X.

It will be found that the only combinations inconsistent with the data are

(α) A truths, B lies, C truths, D truths;

(β) A truths, B lies, C lies, D lies.

For if A is lying, we do not know what B said; and if A and B both truthed, we do not know what C said.

Since (α) and (β) occur respectively twice and eight times out of 81 occasions, D's 27 truths and 54 lies are reduced to 25 truths and 46 lies. The probability is therefore 25/71.

Eddington's solution is certainly correct though dissents continue to be expressed.

IV

I now turn to other facets of Eddington's contributions to classical relativity. It is my judgement that Eddington's greatest contribution to the general theory of relativity is his wondrous treatment of the subject in his *Mathematical Theory of Relativity*. I continue to use it. Besides, the mathematical treatment is interspersed with many striking aphorisms. The one I like best is[25]

Space is not a lot of points close together; it is a lot of distances interlocked.

The following extracts from a review of the book by his 'adversary', Jeans, summarizes very well its distinguishing features:[26]

* More generally, if A, B, C, and D speak the truth with probabilities a, b, c, and d, respectively, then the solution of the same problem, as L. S. Leftwich[24] showed in considerable detail, is

$$\frac{d - acd(1-b)}{1-a(1-b)+a(1-b)(c-2cd+d)}.$$

Everywhere we find indications of ungrudged labour and scrupulous care; we read section after section and each time feel that the matter could not have been better put. Owing to the care which has been expended upon it, the mathematician will read the book with ease and pleasure . . .

The style of the book is admirably clear and concise throughout; we can give it no higher praise than to say that it is fully up to the high standard which Prof. Eddington has led us to expect from him.

Departing from generalities, I should like to dwell briefly on three specific investigations which illustrate Eddington's perception and understanding of classical relativity.

First, most modern students of relativity will be familiar with the fact that the apparent singularity of the Schwarzschild metric at what is now designated as the event horizon is a consequence of the chosen coordinate system and has no further significance. In 1924, Eddington explicitly gave a transformation[27] – now called the Eddington–Finkelstein transformation – which makes this fact manifest. It should, however, be stated that Eddington obtained his transformation for a different purpose and it is not clear from the context that he was addressing himself to the problem of the coordinate singularity.

Second, in view of the current arguments over the validity or otherwise of Einstein's original formula for the rate of emission of gravitational radiation in terms of the varying quadrupole moment of the source, it is interesting to recall that Eddington made,[28] as early as 1922, an explicit *ab initio* calculation of the rate of emission of gravitational energy by a rigid spinning rod and obtained the correct answer, discovering incidentally a numerical error of a factor 2 in Einstein's original formula.

Third, in an investigation published in 1938, Eddington and Clark,[29] independently of Einstein, Infeld, and Hoffmann, considered the N-body problem in general relativity and solved for the metric coefficients in what we should now call the first post-Newtonian approximation. Eddington and Clark did not, however, reduce the problem to the Hamiltonian form nor did they obtain the analogues of the classical ten integrals of the equations of motion. They were primarily concerned with the motion of the centre of mass. The general solution to their problem requires one to obtain the coordinate transformation, consistently in the first post-Newtonian approximation, between any two frames of reference in uniform relative motion i.e., a *post-Galilean transformation.* Eddington and Clark did not obtain such a transformation; but they did solve a restricted problem in the relative motion of two mass-points to which they had addressed themselves.

The foregoing examples, particularly the last one, show that Eddington, if he cared, could formulate and solve problems of depth and complexity in the classical theory of relativity. But he does not seem to have cared much.

V

Of his contributions to classical relativity, Eddington attached the greatest importance to his generalization of Weyl's theory attempting to unify gravitation and electromagnetism. Indeed, in a document[30] discovered among his papers in 1954, Eddington, in an impersonal statement of what he considered as his major scientific accomplishments, gives a prominent place to his generalization of Weyl's theory and 'connected with this, his explanation of the law of gravitation', $G_{\mu\nu} = \Lambda g_{\mu\nu}$, where Λ is the cosmical constant. In any event, the attitude of mind that he formed, in this context at this time, was to crystallize in later years to become a permanent bedrock. For this reason, I shall try to explain briefly the nature of Weyl's theory and Eddington's generalization of it.

At the time Einstein formulated his general theory of relativity, it was thought that the entire physical world required for its description only two fields: the gravitational field and the electromagnetic field. Since Einstein had shown how the gravitational field can be absorbed into the structure of space-time, it was natural that efforts should have been made to absorb the electromagnetic field into the structure of space-time, as well. Clearly, if one is to achieve such a synthesis, one must enlarge the geometrical base of Einstein's theory by a suitable generalization of Riemannian geometry. Weyl and Eddington sought such a generalization in the net effect of displacing a vector, parallel to itself, around a closed infinitesimal contour. In Riemannian geometry, a vector, after describing such a closed contour, is altered in its direction; but is unchanged in its length. Weyl supposed that the length is also changed by an amount proportional to its initial length; and Eddington allowed the change in length to be arbitrary (in the first instance).

In Weyl's theory, consistently with its underlying assumptions, the Christoffel connection, $\Gamma_{ij,k}$ of Riemannian geometry is replaced by

$$\Gamma^*_{ij,k} = \Gamma_{ij,k} + \tfrac{1}{2}(g_{ik}\phi_j + g_{jk}\phi_i - g_{ij}\phi_k), \tag{1}$$

where ϕ_i, $i = 1, 2, 3, 4$, are some smooth functions. Also, in Weyl's theory, we should require that all geometrical relations and physical laws are invariant not only to arbitrary coordinate transformations (as in Einstein's theory) but also to *gauge transformations* i.e., to the substitution

$$\phi_i \to \phi_i - \frac{1}{\lambda}\frac{\partial \lambda}{\partial x^i}, \tag{2}$$

where λ is an arbitrary function. With these postulates, Weyl showed that

$$F_{ik} = \frac{\partial \phi_k}{\partial x^i} - \frac{\partial \phi_i}{\partial x^k}, \tag{3}$$

has all the properties required of the Maxwell tensor; and he achieved the unification of gravitation and electromagnetism in this fashion.

The most important consequence of Weyl's theory, for the theory of gravitation, is that in the absence of an electromagnetic field, Einstein's equation (for the vacuum),

$$G_{ij} = 0, \tag{4}$$

is replaced by

$$G_{ij} = \Lambda g_{ij}, \tag{5}$$

where Λ is a universal constant, and g_{ij} and G_{ij} are, respectively, the metric tensor and the Einstein tensor. The constant Λ in equation (5) is the *cosmical constant* which Einstein had introduced, earlier in 1917, as an afterthought, in order that his theory may allow a static, homogeneous, and isotropic model for the Universe.

Eddington's generalization of Weyl's theory amounts to replacing equation (1) by

$$\Gamma^{*}_{ij,k} = \Gamma_{ij,k} + K_{ik,j} + K_{jk,i} - K_{ij,k}, \tag{6}$$

where $K_{ik,j}$ is some covariant tensor of rank 3 (unspecified at this stage). And again, in the absence of an electromagnetic field, we are led to Einstein's equation including the term in the cosmical constant.

The fact that, by the foregoing equations, we are naturally led to the term in the cosmical constant, convinced Eddington of the *necessity* of including it in Einstein's equation; and it became central to his views. As he explained it,[31]

> *The radius of curvature at any point and in any direction is in constant proportion to the length of a specified material unit placed at the same point and oriented in the same direction.*

Or, conversely,

> *The length of a specified material structure bears a constant ratio to the radius of curvature of the world at the place and in the direction in which it lies.*

Eddington states this central doctrine of his in a variety of ways. Thus,[32]

> *We see that Einstein's law of gravitation is the almost inevitable outcome of the use of material-measuring appliances for surveying the world, whatever may be the actual laws under which material structures are adjusted in equilibrium with the empty space around them.*

Or, again,[33]

> *An electron would not know how large it ought to be unless there existed independent lengths in space for it to measure itself against.*

Indeed, Eddington considered that reverting to Einstein's

equation without the Λ-term is tantamount to reverting to Newtonian theory:[34]

I would as soon think of reverting to Newtonian theory as of dropping the cosmical constant.

This absolute conviction of his resulted in such extreme statements as the following:[35a, 35b]

To set $\Lambda = 0$, implies a reversion to the imperfectly relativistic theory – a step which is no more to be thought of than of a return to the Newtonian theory.

... The position of the cosmical constant seems to me impregnable; and if ever the theory of relativity falls into disrepute the cosmical constant will be the last stronghold to collapse. To drop the cosmical constant would knock the bottom out of space.

Eddington, however, was not alone in his views. I once asked Lemaître, sometime during the late fifties, what, in his judgement, was the most important change wrought by the general theory of relativity in our basic physical concepts. His answer, without a moment's hesitation, was 'the introduction of the cosmical constant!' Similarly, in a letter to Niels Bohr in 1923, Einstein states, unequivocally,[36]

Eddington has come closer to the truth than Weyl.

Indeed, one of the last 'unified field-theories' which Einstein (and, independently, Schrodinger) developed has much in common with Eddington's generalization of Weyl's theory. But Weyl's views were different. He wrote[37] in 1953:

As to Eddington's own creative contributions to the theory, I would say that they consist chiefly of two things: first his idea of an affine field theory, and then his later attempts to explain by epistemological reasons the pure numbers that seem to enter into the constitution of the universe ...

His first contribution certainly has borne fruit. Einstein himself took it up when he formulated an action principle for such a theory (which Eddington, I believe erroneously, had thought unnecessary) ...

... but I am quite sceptical also about Einstein's most recent

unitary field theory. I am pretty sure that the last word on the nature of gravitation is not yet spoken, and I am inclined to believe that it lies in a direction quite different from Eddington's and Einstein's last ideas. The riddle may have to wait a long time for its solution.

In spite of Eddington's positive views, the subsequent history of the cosmical constant has been a chequered one. When Friedmann's cosmological models were found to provide an adequate base for accounting for the simple fact of the Hubble expansion, Einstein and deSitter, in a joint paper, stated that one can do without the cosmical constant. In view of the many exaggerated statements that have been made concerning this supposed 'retraction' of Λ, it is of interest to record precisely what it was they said.[38]

Historically the term containing the 'cosmological constant' Λ was introduced into the field equations in order to enable us to account theoretically for the existence of a finite mean density in a static universe. It now appears that in the dynamical case this end can be reached without the introduction of Λ . . .

. . . The curvature [constant Λ] is, however, essentially determinable, and an increase in the precision of data derived from observations will enable us in the future to fix its sign and to determine its value.

Eddington has written[39] of his meeting with Einstein and of a communication from deSitter, soon after the publication of their paper, which throws an interesting sidelight on this matter (and doubt on the extreme statements that have been attributed to Einstein with respect to his retraction).

Einstein came to stay with me shortly afterwards, and I took him to task about it. He replied: 'I did not think the paper very important myself, but deSitter was keen on it.' Just after Einstein had gone, deSitter wrote to me announcing a visit. He added: 'You will have seen the paper by Einstein and myself. I do not myself consider the result of much importance, but Einstein seemed to think that it was.'

What is the present view, then, on the cosmical constant? One can discern the prevalence of two views: an *extreme view* (expressed, for example, by W. Pauli[40]) that the cosmological term 'is superfluous, not justified, and should be rejected'; and a *moderate view* (expressed, for example, by W. Rindler[41]) that the cosmological term 'belongs to the field equations, much as an additive constant belongs to an indefinite integral'. There is much to recommend the moderate view, since the term in Λ is of no consequence except in the cosmological context and its inclusion hardly increases the complexity of the solutions for the cosmological models one generally considers. In any event, it is clear that no serious student of relativity is likely to subscribe to Eddington's view that 'to set $\Lambda = 0$ is to knock the bottom out of space'.

VI

Let me conclude this part of my consideration of Eddington's work with an anecdote in a lighter vein.

Eddington visited the Physics Department of the University of California at Berkeley in 1924. On that occasion, he shared an office with one Professor W. H. Williams with whom he played golf twice a week at the Claremont Club. On the eve of his departure, the Faculty Club arranged a dinner in Eddington's honour; and Professor Williams was asked to give a speech. As Professor Williams wrote,[42]

> *. . . after some efforts towards solemnity, I descended to doggerel. Eddington, as you know, was an* Alice in Wonderland *fan. This and the allied topsy-turvydom of Carroll and Einstein, together with the irreverent way we both treated the royal game of golf, furnished the motive for the following poem.*

THE EINSTEIN AND THE EDDINGTON

The sun was setting on the links,
The moon looked down serene,
The caddies all had gone to bed,
But still there could be seen
Two players lingering by the trap
That guards the thirteenth green.

The Einstein and the Eddington
Were counting up their score;
The Einstein's card showed ninety-eight
And Eddington's was more.
And both lay bunkered in the trap
And both stood there and swore.

I hate to see, the Einstein said;
Such quantities of sand;
Just why they placed a bunker here
I cannot understand.
If one could smooth this landscape out,
I think it would be grand.

If seven maids with seven mops
Would sweep the fairway clean
I'm sure that I could make this hole
In less than seventeen.
I doubt it, said the Eddington,
Your slice is pretty mean.

Then all the little golf balls came
To see what they were at,
And some of them were tall and thin
And some were short and fat,
A few of them were round and smooth,
But most of them were flat.

The time has come, said Eddington,
To talk of many things:
Of cubes and clocks and meter-sticks
And why a pendulum swings.
And how far space is out of plumb,
And whether time has wings.

I learned at school the apple's fall
To gravity was due.
But now you tell me that the cause
Is merely $G_{\mu\nu}$,
I cannot bring myself to think
That this is really true.

You say that gravitation's force
Is clearly not a pull.
That space is mostly emptiness,
While time is nearly full;
And though I hate to doubt your word,
It sounds a bit like bull.

And space, it has dimensions four,
Instead of only three.
The square on the hypotenuse
Ain't what it used to be.
It grieves me sore, the things you've done
To plane geometry.

You hold that time is badly warped,
That even light is bent:
I think I get the idea there,
If this is what you meant:
The mail the postman brings today,
Tomorrow will be sent.

If I should go to Timbuctoo
With twice the speed of light,
And leave this afternoon at four,
I'd get back home last night.
You've got it now, the Einstein said,
That is precisely right.

But if the planet Mercury
In going round the sun,
Never returns to where it was
Until its course is run,
The things we started out to do
Were better not begun.

And if, before the past is through,
The future intervenes;
Then what's the use of anything;
Of cabbages or queens?
Pray tell me what's the bally use
Of Presidents and Deans.

The shortest line, Einstein replied,
Is not the one that's straight;
It curves around upon itself,
Much like a figure eight,
And if you go too rapidly
You will arrive too late.

But Easter day is Christmas time
And far away is near,
And two and two is more than four
And over there is here.
You may be right, said Eddington,
It seems a trifle queer.

But thank you very, very much,
For troubling to explain;
I hope you will forgive my tears,
My head begins to pain;
I feel the symptoms coming on
Of softening of the brain.

VII

My discussion, so far, has been concerned, principally, with Eddington's contributions to astrophysics and to classical relativity up to the middle twenties. In 1926, when his *Internal Constitution of the Stars* was published, Eddington was forty-four years old. For the following eighteen years, except for occasional excursions into areas of his earlier interests, Eddington was preoccupied with justifying a particular cosmological model of his choice, a model which he further made the basis for his 'fundamental theory unifying quantum theory and relativity theory'. I do not make claims to understand, in any real sense, Eddington's fundamental theory. But there are two premises which, by Eddington's own statements, are basic to his theory. And these two premises, to the extent I can judge, are either not valid or not accepted. But first, I should explain, as objectively as I can, Eddington's perception of the cosmological aspects of the problem.

At the time Eddington wrote his *Mathematical Theory of*

Relativity there were two models of the universe* both dependent on a non-vanishing cosmical constant Λ: Einstein's universe which is static and in hydro-static equilibrium and deSitter's universe which is also static but expanding. Both models are theoretically possible and consistent with the assumptions of homogeneity and isotropy. Since in deSitter's universe there is an expansion, with distant objects receding with increasing velocity, while in Einstein's universe there is no such expansion, Eddington, already in his *Mathematical Theory of Relativity*, chose deSitter's universe as the more likely one, as a model for the astronomical universe, on the strength of some very meagre observational data that were available to him at that time.

Both Einstein's and deSitter's universe are static in the sense that none of the metric coefficients are dependent on time. It was realized later that the static character of deSitter's universe, in spite of the expansion which it exhibits, is derived from the absence of any mass-density in it. Therefore, Einstein's universe was, at that time, the only model of the universe which had no motions in it. As Eddington summed up the situation:[43]

> *Einstein's universe contains matter but no motion; deSitter's universe contains motion but no matter.*

However, in a paper published in 1922, A. Friedmann had shown that Einstein's equation allowed non-static models for the universe which are homogeneous and isotropic. Independently of Friedmann, Lemaître rediscovered the same solutions in 1927 and developed in some detail the astronomical consequences of the theory. Eddington came to know of Lemaître's paper and gave it wide publicity in his writings. Since Λ now became a parameter to which one may assign positive, zero, or negative values, one had a range of cosmological models to choose from. These models continue to provide the basis for comparisons with astronomical observations. But from the outset, Eddington's interests were focused on a particular member of the sequence of possible

* I am excluding the non-static solutions discovered by Friedmann in 1922 since they came to be known generally only several years later.

models. The model he preferred was one in which the universe was initially an Einstein universe with its mass (M) and radius (R_E) related in the manner

$$\Lambda = \frac{1}{R_E} \quad \text{and} \quad \frac{GM}{c^2} = \tfrac{1}{2}\pi R_E; \tag{7}$$

and later by virtue of its instability (a fact which had been demonstrated by Lemaître) the universe began expanding. It could have contracted just as well; but Eddington and others showed that an initial condensation is more likely to lead to an expansion than to a contraction.

Why did Eddington choose this particular model for the description of the astronomical universe? As he explained:[44]

> *I am a detective in search of a criminal – the cosmical constant. I know he exists, but I do not know his appearance; for instance I do not know if he is a little man or a tall man.*

And the way Eddington endeavoured to find the 'appearance' of his 'criminal' is roughly along the following lines.

In Dirac's equation for an electron, moving in the electromagnetic field of a fixed electric charge, the term which involves its mass, m_e, is $m_e c^2/e^2$. Eddington argued that this term arises by virtue of the existence of all the other particles in the universe. More precisely he considered it as a consequence of the 'energy of interchange' with the rest of the charges in the universe suitably averaged; and he satisfied himself that the term must be $\sqrt{N}/R_E$, apart from a possible numerical factor of order unity, where N is the number of particles in the universe and R_E is the radius of the initial static Einstein-universe. In this manner, he obtained the relation,

$$\frac{\sqrt{N}}{R_E} = \frac{m_e c^2}{e^2}. \tag{8}$$

But we also have the relation (*cf.* equation (7))

$$\frac{GNm_p}{c^2} = \tfrac{1}{2}\pi R_E, \tag{9}$$

where m_p denotes the mass of the proton. From these two relations, we find

$$\mathcal{N} = \frac{\pi^2}{4} \frac{e^4}{(Gm_p m_e)^2} = 1.28 \times 10^{79}$$

and

$$\frac{1}{\Lambda} = R_E = \tfrac{1}{2}\pi \frac{e^4}{Gm_p m_e^2 c^2} = 1.07 \times 10^9 \text{ light years.} \quad (10)$$

Eddington considered these values to be sufficiently in accord with the observations that he felt that he had detected his 'criminal'. From this point on, he had no doubts about the soundness of his arguments. Indeed, in a course of lectures he gave at the Institute for Advanced Studies in Dublin in 1943, he stated[45]

> *At no time during the past sixteen years have I felt any doubt about the correctness of my theory.*

It is now pertinent to ask what the present status of Eddington's cosmological model is. The advances in our knowledge since 1944 leave little doubt that it is not the right one. The evidences derive, principally, from the discovery of the universal 3 K black-body radiation and the fact that helium is of primordial origin.

The fact that the universe is pervaded by a homogeneous isotropic radiation-field with a Planckian distribution at a temperature of 3 K, implies that the universe has expanded by a factor of some 1300 from the time when the temperature of the universal radiation was 4000 K and the decoupling of the matter and the radiation occurred. Similarly, the production of primordial helium by nucleosynthesis implies that the universe, at one time, was in a state in which the density was of the order of 1000 g/cm^3 and the temperature was of the order of 10^9 K. These requirements, in turn, imply that the radius of the universe, at that time, must have been smaller by a factor of the order of 10^9. Variations in the radius of the universe of this magnitude are simply incompatible with Eddington's cosmological model.

On June 9, 1944, barely five months before he died, Eddington, in his last appearance at a meeting of the Royal

Astronomical Society, presented a paper on 'The Recession-Constant of the Galaxies'. On this occasion, Eddington stated that Λ given by equation (10)[46]

> is *the cosmical constant*,

and further that

> . . . *the time scale for the evolution of the universe is definitely less than* 90×10^9 *years, and I do not see much prospect of evading this limit.*

After the presentation of the paper, the following exchange between Eddington and G. C. McVittie took place:[47]

> Dr. McVittie. *The theory seems to be based entirely on the model which starts expanding from the Einstein universe. It is known that the observations of the distribution of spiral nebulae 'in depth' provide a very delicate test for distinguishing between one model universe and another. What becomes of the theory if observation does not happen to select the particular model you have used?*
>
> Sir Arthur Eddington. *I do not think we shall have data accurate enough to settle this point for a very long time, so that I do not think I need consider the contingency! We have to try to reach a conclusion by other means than a direct comparison with observation.*

The contingency to which Eddington refers has occurred; and McVittie's question remains unanswered.

VIII

I have already described Eddington's line of argument in relating the number of particles in the universe with atomic constants. For the further development of his theory, he needed to enumerate the number of cells of volume h^3 in the available phase-space of the initial static Einstein-universe. The required enumeration was, in Eddington's mind, closely related to, if not the same as, the enumeration one makes when deriving the equation of state of a degenerate electron-gas consistently with Pauli's exclusion principle. Since I am involved in this matter, I shall read from a detailed statement that Eddington has made of the evolution of his ideas in this regard, in an address to the Tercentenary Conference of Arts and Sciences at Harvard University,[48] in the summer of 1936:

. . . in the stars the temperature of 10 million degrees causes most of the satellite electrons to be torn away from the atom, and what is left of the atom is a tiny structure. The atoms or ions are so reduced in size that they will not jam until densities 100 000 times greater are reached. For this reason, the perfect gas state continues up to much higher densities in the stars. The sun and other dense stars insisted on obeying the theory worked out for a perfect gas, as they had every right to do, since their material was perfect gas.

There was, therefore, nothing to prevent stellar matter from becoming compressed to exceedingly high density; and it suggested itself that the densities which had been calculated from observation for certain stars called white dwarfs, which had seemed impossibly high, might be genuine after all.

In reaching this conclusion I was not without a certain misgiving. I was uneasy as to what would ultimately happen to these superdense stars. The star seemed to have got itself into an awkward fix. Ultimately its store of subatomic energy would give out and the star would then want to cool down. But could it? The enormous density was made possible by the high temperature which shattered the atoms. If the material cooled it would presumably revert to terrestrial density. But that meant that the star must expand to say 5 000 times its present bulk. But the expansion requires energy – doing work against gravity; and the star appeared to have no store of energy available. What on earth was the star to do if it was continually losing heat, but had not enough energy to get cold!

The high density of the companion of Sirius was duly confirmed by Professor Adams – but this puzzle remained. Shortly afterward Prof. R. H. Fowler came to the rescue in a famous paper, in which he applied a new result in wave mechanics which had just been discovered. It is a remarkable coincidence that just at the time when matter of transcendently great density was discovered in astronomy, the mathematical physicists were quite independently turning attention to the same subject. I suppose that up to 1924 no one had given a serious thought to abnormally dense matter; but just when it cropped up in astronomy it cropped up in physics as well. Fowler showed that the newly discovered Fermi–Dirac statistics saved the star from the unfortunate fate which I had feared.

Not content with letting well alone, physicists began to improve on Fowler's formula. They pointed out that in white dwarf conditions the electrons would have speeds approaching the velocity of light, and there would be certain relativity effects which Fowler had neglected. Consequently, Fowler's formula, called the ordinary *degeneracy formula, came to be superseded by a newer formula, called the* relativistic *degeneracy formula. All seemed well until certain researches by Chandrasekhar brought out the fact that the relativistic formula put the stars back in precisely the same difficulty from which Fowler had rescued them. The small stars could cool down all right, and end their days as dark stars in a reasonable way. But above a certain critical mass (two or three times that of the sun) the star could never cool down, but must go on radiating and contracting until heaven knows what becomes of it. That did not worry Chandrasekhar; he seemed to like the stars to behave that way, and believes that that is what really happens. But I felt the same objections as 12 years earlier to this stellar buffoonery; at least it was sufficiently strange to rouse my suspicion that there must be something wrong with the physical formula used.*

I examined the formula – the so-called relativistic degeneracy formula – and the conclusion I came to was that it was the result of a combination of relativity theory with a nonrelativistic quantum theory. I do not regard the offspring of such a union as born in lawful wedlock. The relativistic degeneracy formula – the formula currently used – is in fact baseless; and, perhaps rather surprisingly, the formula derived by a correct application of relativity theory is the ordinary formula – Fowler's original formula which every one had abandoned. I was not surprised to find that in announcing these conclusions I had put my foot in a hornet's nest; and I have had the physicists buzzing about my ears – but I don't think I have been stung yet. Anyhow, for the purposes of this lecture, I will assume that I haven't dropped a brick.

I venture to refer to a personal aspect of this investigation, since it shows how closely different branches of science are interlocked.

At the time when my suspicion of the relativistic degeneracy formula was roused by Chandrasekhar's results, it was very inconvenient to me to spare time to follow it up, because I was immersed in a long investigation in a different field of thought. This work, which had occupied me for six years, was nearing completion and there remained only one problem, namely, the accurate theoretical calculation of the cosmical constant, needed to round it off. But there I had completely stuck. I had, however, secured a period of four months free from distractions which I intended to devote to it – to make a supreme effort, so to speak. But having incautiously begun to think about the degeneracy formula I could not get away from it. It took up my time. The months slipped away, and I had done nothing with the problem of the cosmical constant. Then one day in trying to test my degeneracy results from all points of view, I found that in one limiting case it merged into a cosmical problem. It gave a new approach to the very problem which I had had to put aside – and from this new approach the problem was soluble without much difficulty. I can see now that it would have been very difficult to get at it in any other way; and it is most unlikely that I should have made any progress if I had spent the four months on the direct line of attack which I had planned.

The paper which I read to the mathematical section a few days ago, giving a calculation of the speed of recession of the spiral nebulae and the number of particles in the universe, had an astronomical origin. It was not, however, suggested by consideration of the spiral nebulae. It arose out of the study of the companion of Sirius and other white dwarf stars.

Let me explain what the points at issue were. Fowler's discussion of the state of matter in the white-dwarf stars was based on precisely the same theory of the degenerate electron-gas that had been made familiar by Sommerfeld's electron theory of metals. The equation of state governing such an electron gas is

$$p = \frac{1}{20}\left(\frac{3}{\pi}\right)^{\frac{2}{3}} \frac{h^2}{m_e} n^{\frac{5}{3}}, \tag{11}$$

where p denotes the pressure and n the number of electrons per cm³. However, at the densities prevailing at the centres of white-dwarf stars, the electrons at the Fermi threshold begin to have velocities comparable to that of light. If one allows for this circumstance, in a way that was common then and continues to be common to this day, one finds that the equation of state departs from that given by equation (11) and, in the limit of very high electron-concentrations, tends to

$$p = \frac{1}{8}\left(\frac{3}{\pi}\right)^{\frac{1}{3}} hc\, n^{\frac{4}{3}} \quad (n \to \infty). \tag{12}$$

It is this modification of the equation of state which Eddington considered as 'baseless'.

The consequences which result from using the equation of state in its non-relativistic form (11) and in its exact form with the low- and the high-density limits (11) and (12), are the following.

On the basis of the non-relativistic equation of state (11), one finds that stellar masses, in equilibrium, have radii which vary inversely as the cube root of the mass. Finite equilibrium configurations are, therefore, possible for all masses. It is this fact which Eddington considered as eminently satisfactory. However, when one uses the exact form of the equation of state, with its high-density limit (12), one finds that no equilibrium state is possible if the mass exceeds the limit,

$$M_{\text{limit}} = 0.197\left(\frac{hc}{G}\right)^{\frac{3}{2}} \frac{1}{(\mu_e H)^2} = 5.76\,\mu_e^{-2}\odot, \tag{13}$$

where μ_e is the mean molecular weight per electron. The complete mass-radius relation[49] that one obtains is illustrated in Fig. 4.

It is this fact, that no finite degenerate stellar-configurations exist for masses exceeding the limit (13), which Eddington considered as 'stellar buffoonery'. As he stated in an earlier context,[50]

> *Chandrasekhar, using the relativistic formula which has been accepted for the last five years, shows that a star of mass greater than a certain limit 𝔐 remains a perfect gas and can never cool down. The star has to go on radiating and radiating, and contracting and contracting until, I suppose, it gets down to a few km*

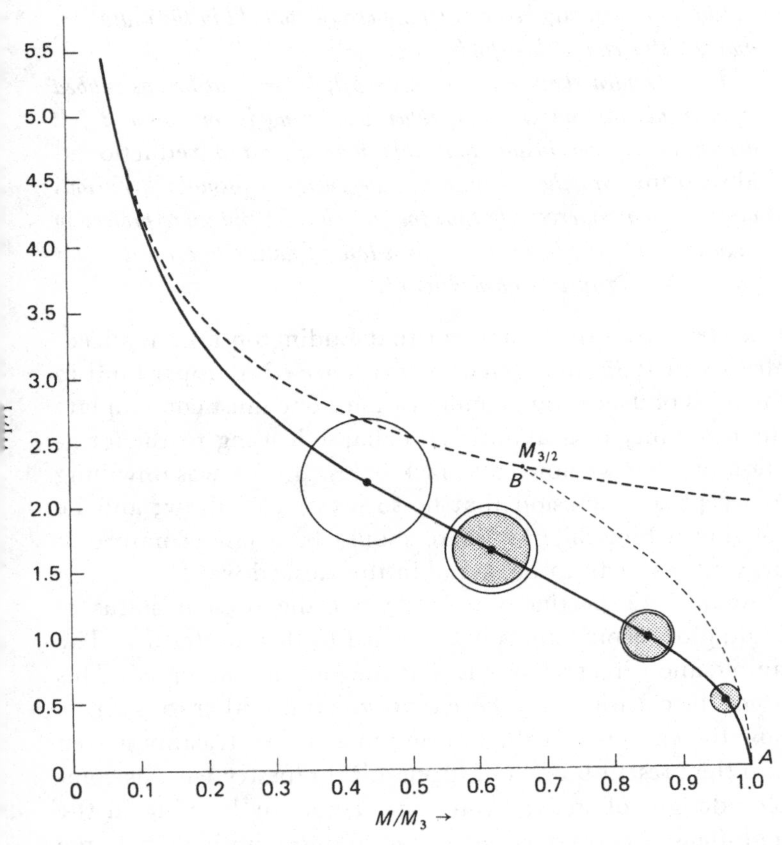

Fig. 4. The full-line curve represents the exact (mass–radius) relation for completely degenerate configurations. The mass, along the abscissa, is measured in units of the limiting mass (denoted by M_3) and the radius, along the ordinate, is measured in the unit $l_1 = 7.72\mu^{-1} \times 10^8$ cm. The dashed curve represents the relation that follows from the equation of state (11); at the point B along this curve, the threshold momentum p_0 of the electrons at the centre of the configuration is exactly equal to $m_e c$. Along the exact curve, at the point where a full circle (with no shaded part) is drawn, p_0 (at the centre) is again equal to $m_e c$; the shaded parts of the other circles represent the regions in these configurations where the electrons may be considered to be relativistic ($p_0 > m_e c$). (Reproduced from S. Chandrasekhar, *Mon. Not. Roy. Astr. Soc.*, **95**, no. 219, 1935.)

radius, when gravity becomes strong enough to hold in the radiation, and the star can at last find peace.

Dr. Chandrasekhar had got this result before, but he has rubbed it in in his last paper; and, when discussing it with him, I felt driven to the conclusion that this was almost a reductio ad absurdum *of the relativistic degeneracy formula. Various accidents may intervene to save the star, but I want more protection than that. I think there should be a law of nature to prevent a star from behaving in this absurd way!*

It is clear from this statement that Eddington fully realized, already in 1935, that given the existence of an upper limit to the mass of degenerate configurations, one must contemplate the possibility of gravitational collapse leading to the formation of what we now call black holes. But he was unwilling to accept a conclusion that he so presciently drew; and he convinced himself that 'there should be a law of nature to prevent a star from behaving in this absurd way!'

Again it is pertinent to ask what the present status of Eddington's convictions with regard to this matter are. The simple and direct answer is that they are not accepted. This is not the occasion and there is not the time either to describe how the existence of the limiting mass is inextricably woven into the present fabric of astronomical tapestry with its complex designs of stellar evolution, of nuclear burning in the high-density cores of certain stars, and gravitational collapse leading to the supernova phenomenon and the formation of neutron stars of nearly the same mass and of black holes. All these are discernible even to the most casual observer. For my part I shall only say that I find it hard to understand why Eddington, who was one of the earliest and staunchest supporters of the general theory of relativity, should have found the conclusion that black holes may form during the natural course of the evolution of the stars, so unacceptable.

IX

It would appear, then, that two of the main pillars of the grand edifice that Eddington built in his 'fundamental theory' have collapsed. Forgetting this fact for a moment, what are we to make of the structure itself? Here are two views.[51,52]

> *Eddington's work, if correct, is extremely important, but most of those who have tried to read his work have not been able to agree with his conclusions. His papers are very clear up to a point and then at the critical moment they become obscure, to become clear again after the important results have been deduced. There is certainly no logical deduction of the conclusions from explicitly stated axioms and hypotheses, and Eddington himself was aware of this. Once after a long discussion with the present writer, which achieved very little, Eddington said: 'I can't quite see through the proof, but I am sure the result is correct.'* (*A. H. Wilson*)
>
> *Eddington's 'unified theory', apart from its obscurity, explains too much; indeed it explains everything, and hitherto such theories have generally been found ultimately to explain nothing. At best, it is a fragmentary work, which contains flashes of insight that will be appreciated by future generations, like Leonardo da Vinci's scientific researches, probably after what is significant in them has been more completely discovered by very different paths and methods.* (*J. G. Crowther*)

There can, however, be little doubt that, even though Eddington's edifice may, in part, be in ruins, there are high columns that are still standing erect. One example must suffice. In his treatment of Dirac's equation, Eddington developed a calculus of *E*-numbers which is, basically, the group algebra of sixteen elements which satisfy the anti-commutation rules of the Dirac matrices. This development represents a considerable achievement in itself; and it has had important repercussions in the current renewed interest in Clifford algebras. The basic elements of Eddington's developments are the following.

Eddington starts by defining five E-numbers by the representation,

$$E_1 = \begin{vmatrix} i\sigma_1 & 0 \\ 0 & i\sigma_1 \end{vmatrix}, \quad E_2 = \begin{vmatrix} i\sigma_3 & 0 \\ 0 & i\sigma_3 \end{vmatrix}, \quad E_3 = \begin{vmatrix} 0 & -\sigma_2 \\ \sigma_2 & 0 \end{vmatrix},$$

$$E_4 = \begin{vmatrix} i\sigma_2 & 0 \\ 0 & -i\sigma_2 \end{vmatrix}, \quad \text{and} \quad E_5 = \begin{vmatrix} 0 & i\sigma_2 \\ i\sigma_2 & 0 \end{vmatrix}, \tag{14}$$

where σ_1, σ_2, and σ_3 are the Pauli 2×2-matrices. These E-numbers satisfy the commutation rules,

$$E_\mu E_\nu + E_\nu E_\mu = -2\delta_{\mu\nu} \quad (\mu, \nu = 1, \ldots, 5), \tag{15}$$

and

$$E_1 E_2 E_3 E_4 = iE_5. \tag{16}$$

And the sixteen elements of Eddington's algebra are

$$i, E_\mu \quad \text{and} \quad E_\mu E_\nu \quad (\mu, \nu = 1, \ldots, 5; \mu \neq \nu). \tag{17}$$

Eddington went further to define the square of the E-algebra as 16×16 complex matrices. This 'double E-frame' corresponds to one of three Clifford algebras in a nine-dimensional base space. Since recent work on super-symmetric gauge-fields is based on eight and nine-dimensional Clifford algebras, Eddington was, in this respect, very much ahead of his time.

Eddington attached considerable importance to finding idempotent E-numbers i.e., elements E of the algebra satisfying $E^2 = E$. These are the projection operators for momentum and spin in quantum electrodynamics. And finally, his realization that the E-algebra is *five*-dimensional over the real numbers led him to introduce (for the first time in particle physics) the notion of 'chirality' derived from the choices, $+i$ or $-i$, on the right-hand side of equation (16). It should also be stated that Eddington was the first to identify the algebra of the 4×4 real matrices and its significance. This algebra was later discovered by E. Majorana;

and in the physics literature it is often referred to as the 'Majorana spinors'.*

X

I should like to conclude by briefly tracing through Eddington's writings his changing attitude to scientific research. When we juxtapose these attitudes with the directions of his research, we can, perhaps, catch some glimpse of the sources of his strength and of his weakness.

In his address on 'The Internal Constitution of the Stars' to the British Association in 1920 (from which I have quoted earlier), Eddington discussed at some length the place of speculation and of idealized models in scientific inquiry. This is what he said:[53]

> *. . . I wonder what is the touchstone by which we may test the legitimate development of scientific theory and reject the idly speculative. We all know of theories which the scientific mind instinctively rejects as fruitless guesses; but it is difficult to specify their exact defect or to supply a rule which will show us when we ourselves do err. It is often supposed that to speculate and to make hypotheses are the same thing; but more often they are opposed. It is when we let our thoughts stray outside venerable, but sometimes insecure, hypothesis that we are said to speculate. Hypothesis limits speculation. Moreover, distrust of speculation often serves as a cover for loose thinking – wild ideas take anchorage in our minds and influence our outlook; whilst it is considered too speculative to subject them to the scientific scrutiny which would exorcise them.*
>
> *If we are not content with the dull accumulation of experimental facts, if we make any deductions or generalizations, if we seek for any theory to guide us, some degree of speculation cannot be avoided. Some will prefer to take the interpretation which seems to be most immediately indicated and at once adopt that as an hypothesis; others will rather seek to explore and classify the widest possibilities which are not definitely inconsistent with the facts. Either choice*

* I am greatly indebted to Dr. N. Salingaros for drawing my attention to the significance and the importance of Eddington's work on his *E*-numbers.

has its dangers: the first may be too narrow a view and lead progress into a cul-de-sac; the second may be so broad that it is useless as a guide and diverges indefinitely from experimental knowledge. When this last case happens, it must be concluded that the knowledge is not yet ripe for theoretical treatment and speculation is premature. The time when speculative theory and observational research may profitably go hand in hand is when the possibilities – or, at any rate, the probabilities – can be narrowed down by experiment, and the theory can indicate the tests by which the remaining wrong paths may be blocked up one by one.

The mathematical physicist is in a position of peculiar difficulty. He may work out the behaviour of an ideal model of material with specifically defined properties, obeying mathematically exact laws, and so far his work is unimpeachable. It is no more speculative than the binomial theorem. But when he claims a serious interest for his toy, when he suggests that his model is like something going on in nature, he inevitably begins to speculate. Is the actual body really like the ideal model? May not other unknown conditions intervene? He cannot be sure, but he cannot suppress the comparison; for it is by looking continually to nature that he is guided in his choice of a subject. A common fault, to which he must often plead guilty, is to use for the comparison data over which the more experienced observer shakes his head; they are too insecure to build extensively upon. Yet even in this, theory may help observation by showing the kind of data which it is especially important to improve.

I think that the more idle kinds of speculation will be avoided if the investigation is conducted from the right point of view. When the properties of an ideal model have been worked out by rigorous mathematics, all the underlying assumptions being clearly understood, then it becomes possible to say that such and such properties and laws lead precisely to such and such effects. If any other disregarded factors are present, they should now betray themselves when a comparison is made with nature. There is no need for disappointment at the failure of the model to give perfect agreement with observation; it has served its purpose, for it has distinguished what are the features of the actual phenomena which require new conditions for their explanation. A general preliminary agreement with observation is necessary, otherwise the model is hopeless; not that it is necessarily wrong so far as it goes, but it has evidently put the less essential properties foremost. We have been pulling at the wrong end of the tangle, which has to be unravelled by a different approach. But after a general agreement with observation is established, and the tangle begins to loosen, we should always make ready for the next knot. I suppose that the applied mathematician whose theory has just passed one still more stringent test by observation ought not to feel satisfaction, but rather disappointment – 'Foiled again! This time I had *hoped to find a discordance which would throw light on the points where my model could be improved.' Perhaps that is a counsel of perfection; I own that I have never felt very keenly a disappointment of this kind.*

Our model of nature should not be like a building – a handsome structure for the populace to admire, until in the course of time someone takes away a corner-stone and the edifice comes toppling down. It should be like an engine with movable parts. We need not fix the position of any one lever – that is to be adjusted from time to time as the latest observations indicate. The aim of the theorist is to know the train of wheels which the lever sets in motion – that binding of the parts which is the soul of the engine.

There is hardly anything in the foregoing statement to which any serious practitioner of theoretical astrophysics will object. Eddington concluded a lecture on 'The Source of Stellar Energy' given at this time with the modest appraisal:[54]

I should have liked to have closed these lectures by leading up to some great climax. But perhaps it is more in accordance with the true conditions of scientific progress that they should fizzle out with a glimpse of the obscurity which marks the frontiers of present knowledge. I do not apologize for the lameness of the conclusion, for it is not a conclusion. I wish I could feel confident that it is even a beginning.

And this is entirely consistent with the approach to scientific inquiry that I have read. A shift in attitude is already discernible in this statement made two years later:[55]

In science we sometimes have convictions as to the right solution of a problem which we cherish but cannot justify; we are influenced by some innate sense of the fitness of things.

Soon Eddington became cocksure of his views on the cosmical constant, on his cosmological model, on relativistic degeneracy, on the formation of black holes, and, indeed, on his whole approach to 'the unification of quantum theory and relativity theory'. This is abundantly clear from the various quotations from his writings and his lectures that I have read in these contexts. This radical shift in Eddington's attitude is strikingly illustrated by contrasting the modest appraisal of his work on the internal constitution of the stars in 1926 with the self-assured confidence of the remark that he made to me ten years later:

> *. . . You look at it from the point of view of the star; I look at it from the point of view of nature.*

Clearly, at this time, Eddington's views no longer had the kind of sureness which was not cocksureness.

In spite of his expressed confidence in the correctness of his fundamental theory, Eddington must have been deeply frustrated by the neglect of his work by his contemporaries. This frustration is expressed in his plaintive letter to Dingle late in 1944:[56]

> *I am continually trying to find out why people find the procedure obscure. But I would point out that even Einstein was considered obscure, and hundreds of people have thought it necessary to explain him. I cannot seriously believe that I ever attain the obscurity that Dirac does. But in the case of Einstein and Dirac people have thought it worthwhile to penetrate the obscurity. I believe they will understand me all right when they realize they have got to do so – and when it becomes the fashion 'to explain Eddington'.*

And it is haunting to read[57]

> *In his last years, his ghostly pale face was drawn with suffering as he sat in his long reveries.*

Perhaps Eddington foresaw the direction of his future scientific inquiries when he narrated the story of Daedalus and Icarus in his British Association Address of 1920:[58]

> *In ancient days two aviators procured to themselves wings. Daedalus flew safely through the middle air across the sea, and was*

duly honoured on his landing. Young Icarus soared upwards towards the Sun till the wax melted which bound his wings, and his flight ended in fiasco. In weighing their achievements perhaps there is something to be said for Icarus. The classic authorities tell us that he was only 'doing a stunt', but I like to think of him as the man who certainly brought to light a constructional defect in the flying machines of his day. So, too, in Science, cautious Daedalus will apply his theories where he feels most confident they will safely go; but by his excess of caution their hidden weaknesses cannot be brought to light. Icarus will strain his theories to the breaking-point till the weak joints gape. For a spectacular stunt? Perhaps partly; he is often very human. But if he is not yet destined to reach the Sun and solve for all time the riddle of its constitution, yet we may hope to learn from his journey some hints to build a better machine.

And so today, we remember with reverence a great spirit that soared undaunted towards the Sun.

References

Books by Eddington and the abbreviations by which they are referred to in the References

1914 *Stellar Movements and the Structure of the Universe* (London: Macmillan & Co.).

1915 *Report on the Relativity Theory of Gravitation* (London: Physical Society).

1920 *Space, Time and Gravitation* (Cambridge: University Press). *S.T.G.*

1923 *Mathematical Theory of Relativity* (Cambridge: University Press). *M.T.R.*

1926 *Internal Constitution of the Stars* (Cambridge: University Press). *I.C.S.*

1927 *Stars and Atoms* (Oxford: Clarendon Press). *S.A.*

1928 *The Nature of the Physical World* (Cambridge: University Press). *N.P.W.*

1929 *Science and the Unseen World* (London: Allen & Unwin). *S.U.W.*

1933 *The Expanding Universe* (Cambridge: University Press). *E.U.*

1935 *New Pathways in Science* (Cambridge: University Press). *N.P.S.*

1936 *Relativity Theory of Protons and Electrons* (Cambridge: University Press).

1946 *Fundamental Theory* (Cambridge: University Press).

Other books referred to are:

A. Vibert Douglas, *The Life of Arthur Stanley Eddington* (London: Thomas Nelson & Sons, 1957). *V.D.*

J. G. Crowther, *British Scientists of the Twentieth Century*, Chap. IV, (London: Routledge & Kegan Paul, 1952). *J.G.C.*

A. S. Eddington, Forty years of astronomy. In *Background to Modern Science*, J. Needham and W. Pagel, Eds. (Cambridge: University Press, 1938). *J.N. & W.P.*

References

1. *Astrophys. J.*, **101**, 133 (1945).
2. *V.D.*, 103.
3. *J.G.C.*, 177, or *S.A.*, 24.
4. *N.P.S.*, 207.
5. *Ibid.*, 170.
6. *S.U.W.*, 33.
7. *Ibid.*, 54–6.
8. *J.N. & W.P.*, 120–1.
9. *I.C.S.*, 15–16; 245.
10. *Observatory*, **43**, 353–5 (1920).
11. *Nature*, May 1, 1926 (Supp.), No. 2948, p. 30.
12. *Astrophys. J.*, **101**, 134 (1945).
13. *Observatory*, **52**, 349 (1929).
14. *Ibid.*, **49**, 250 and 335 (1926).
15. *Ibid.*, **40**, 424–6 (1917).
16. *Notes & Records Roy. Soc. Lond.*, **30**, 249 (1976).

17*a*. *S.T.G.*, 113–14.

17*b*. *J.N. & W.P.*, 140–2.

18. *Mon. Not. R. Astr. Soc.*, **85**, 672 (1924–5).
19. *Observatory*, **42**, 389–98 (1919).
20. *Ibid.*, **55**, 5 (1932).
21. *Ibid.*, **46**, 142 (1923).
22. *N.P.S.*, 121.
23. *Math. Gazette*, **19**, 256–7 (1935).
24. *Ibid.*, **20**, 309–10 (1936).
25. *M.T.R.*, 10.
26. *Observatory*, **46**, 193 (1923).
27. *Nature*, **113**, 192 (1924).
28. *Proc. Roy. Soc.* (*Lond.*) A, **102**, 268 (1922).
29. *Ibid.*, **166**, 465 (1938).
30. *V.D.*, 189–92.
31. *M.T.R.*, 153.
32. *Ibid.*, 154.
33. *Ibid.*, 155.
34. *E.U.*, 35.

35*a*. *N.P.S.*, 315.

35*b*. *E.U.*, 147–8.

36. *Einstein: A Centenary Volume*, Ed. A. P. French (Cambridge, Mass.: Harvard University Press, 1979), p. 274.
37. *V.D.*, 57.
38. *Proc. Nat. Acad. Sci.*, **18**, 213 (1932).
39. *J.N. & W.P.*, 128.
40. W. Pauli, *Theory of Relativity* (London: Pergamon Press, 1958), Translated by G. Field, p. 220.

41. W. Rindler, *Essential Relativity* (Berlin: Springer-Verlag, 1977), p. 226.
42. *Records of R.A.S. Club* 1925–1953, Ed. G. J. Whitrow, p. xxiv–xxvii.
43. *Proc. Phys. Soc.*, **44**, 6 (1932).
44. *E.U.*, 87.
45. *Dublin Inst. Adv. Studies* A, **2**, 1 (1943).
46. *Observatory*, **65**, 211 (1944).
47. *Ibid.*, 212.
48. *Ann. Rep. Smithsonian Institution* (Washington, D.C.: U.S. Govt. Printing Office, 1938), pp. 137–9.
49. *Mon. Not. R. Astr. Soc.*, **95**, 207 (1935).
50. *Observatory*, **58**, 37 (1935).
51. *Cambridge Rev.*, **66**, 171 (1945).
52. *J.G.C.*, 195.
53. *Observatory*, **43**, 356–7.
54. *Nature*, May 1, 1926 (Supp.), No. 2948, p. 32.
55. *N.P.W.*, 337.
56. *J.G.C.*, 194.
57. *Ibid.*, 143.
58. *Observatory*, **43**, 357–8 (1920).

Notes on the Photographs (Figs. 2 and 3)

Fig. 2 (page 9): The photograph reproduced in this figure is an enlargement of one taken by J. L. Dreyer at the first meeting of the International Astronomical Union after the First World War, in Rome, May, 1922. The original was presented to the author by H. W. Newton (formerly Chief Assistant at the Royal Greenwich Observatory) in November of 1953.

Fig. 3. (page 23). This photograph is a reproduction of one presented to the author by J. H. Oort (formerly Director of the Leiden Observatory, Holland) in 1953 with the note that it 'was taken by deSitter's oldest son (now professor of geology in Leiden) on September 26, 1923. The picture was taken in deSitter's study... We recently found the original negative.'

Index